Tower Bridge

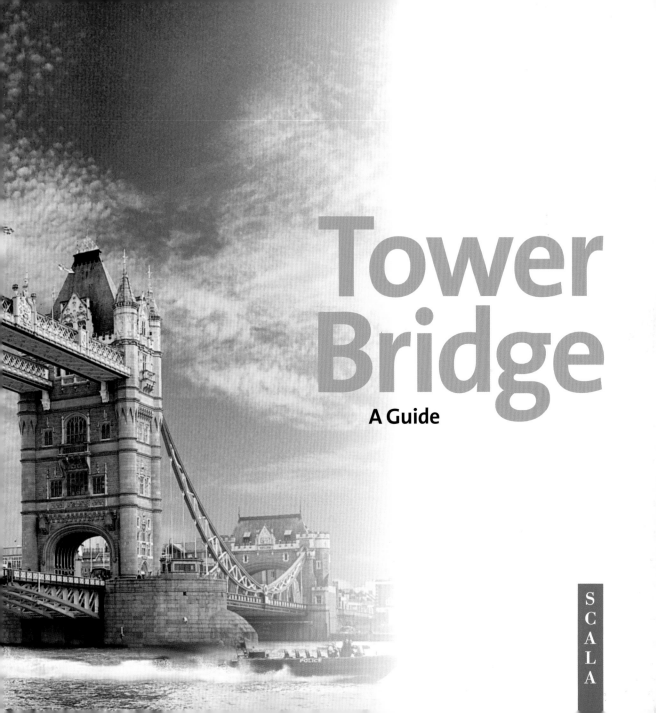

Tower Bridge

A Guide

SCALA

'We are an **old people** but a young nation. Our **trade** is young, our **engineering** is young.'

SAMUEL SMILES, *LIVES OF THE ENGINEERS*, 1861

Tower Bridge is one of the most famous bridges in the world. It is an icon of London and a symbol of the sheer ingenuity of Victorian engineering. When it was built in the 1880s London was the largest and most modern city in the world, a centre of commerce and industry, and the hub of an empire that at its height spanned a fifth of the globe.

Back then, the Thames was still London's lifeblood, as it had been since the city's beginnings. From Roman times until the Middle Ages, the city's port was contained in an area called the Pool of London between London Bridge and Tower Hill. The first stone London Bridge, spiky-backed with houses and later traitors' heads on spikes, was erected in the 12th century. It lasted for nearly 600 years and acted as the western barrier to all seagoing vessels travelling up the Thames.

RIGHT: View of the Tower of London and London Bridge looking northwest, 1766.

A View of the Tower with the Bridge & part of the City of London from the River.

Vue de la Tour, du Pont et partie de la Ville de Londres prise de la Rivière.

268 metres

Length of the bridge

As London expanded in the 18th century, new river crossings were added to the west. Meanwhile, to the east, the Pool of London grew as Britain emerged as a major maritime trading power. Where the loading and unloading of ships had previously taken place on the quayside, with small boats ferrying cargos ashore, now purpose-built docks, wharves and warehouses were established.

The Pool of London in silhouette by L. Hummel, 1926.

Tower Bridge from Custom House Quay by Edwin Smith, c.1957.

Crossing the river

By the 1870s a million people were living east of London Bridge. Their only roadway across the Thames remained that bridge, which was used by 128,000 pedestrians and 20,500 vehicles almost every day. Getting across it could take hours. By comparison, the 2.3 million people who lived to the west of London Bridge had 12 bridges up to Hammersmith at their disposal.

There were always the 'wherries' – the small boats rowed by the Thames watermen. Like modern black cabs, the watermen had been pivotal to city's transport network for centuries.

In 1843 a more modern and less damp alternative arrived when the first tunnel under the Thames between Wapping and Rotherhithe opened. Built by Marc Brunel and his son Isambard Kingdom Brunel, it was a remarkable feat of engineering and christened the 'Eighth Wonder of the World' by the press. However, it was only suitable for pedestrians and the novelty soon wore off.

The need for a new river crossing did not. In 1872 a bill was put before Parliament calling for a new 'Tower Bridge' to be built in east London. But any agreement on its design, or parliamentary approval for its construction, would not come until 1885.

Opening dates of London's Bridges

Putney	1729
Westminster	1750
Blackfriars	1759
Battersea	1771
Vauxhall	1817
Waterloo	1817
Southwark	1819
Hammersmith	1827
New London Bridge	1831
Hungerford	1845
Chelsea	1858
Lambeth	1861
Albert Bridge	1871
Wandsworth	1873

OPPOSITE, ABOVE: View of the Thames, looking east down the river towards Tower Bridge, 1901.

OPPOSITE, BELOW: Traffic flows again after being stopped for the raising of the bridge, London, 1914.

LEFT: A Billingsgate fish porter by the docks with crates of fish balanced on his head, c.1918.

TOWER BRIDGE
OVER THE RIVER THAMES.
ENGINEERS – PETER WILLIAM BARLOW AND ROBERT RICHARDSON
1862

The competition

ABOVE: Proposed suspension bridge across the River Thames, 1862. The approaches were modelled on the Leaning Tower of Pisa.

The design of Tower Bridge, like many of London's other great river crossings, from the new London Bridge of 1831 right up to the Millennium Bridge in 1996, was chosen by public competition.

In 1876 the Corporation of London invited amateur and professional engineers 'to submit a Design or Designs for a Low-Level Bridge with mechanical opening or openings'. All the proposals had to meet strict conditions. If London was a modern metropolis of steam engines, its rivers and roads continued to be the preserve of sailing ships and horse-drawn carriages and carts. Accordingly, any new bridge had to provide a clearance of 43.5 metres (143 feet) for tall-masted sailing ships.

Its approaches and roadways, meanwhile, had to be level enough not to tire horses. Humped bridges and the price of hay and oats were as vital concerns then as the cost of petrol and the congestion charge are now.

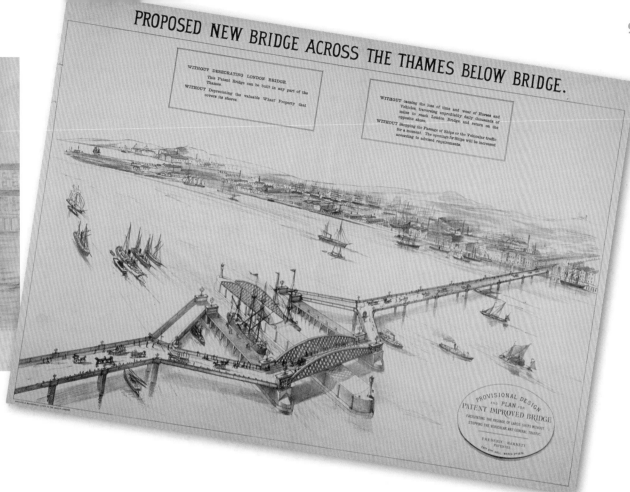

PROPOSED NEW BRIDGE ACROSS THE THAMES BELOW BRIDGE.

WITHOUT DESECRATING LONDON BRIDGE.
This Patent Bridge can be built in any part of the Thames.
WITHOUT Depreciating the valuable Wharf Property that covers its shores.

WITHOUT causing the loss of time and wear of Horses and Vehicles, traversing unprofitably daily thousands of miles to reach London Bridge, and return on the opposite shore.
WITHOUT Stopping the Passage of Ships or the Vehicular traffic for a moment. The openings for Ships will be increased according to advised requirements.

PROVISIONAL DESIGN and PLAN for PATENT IMPROVED BRIDGE FACILITATING THE PASSAGE OF LARGE SHIPS WITHOUT STOPPING THE VEHICULAR AND GENERAL TRAFFIC
FREDERIC BARNETT PATENTEE

Around 50 schemes were submitted for consideration. Mr F.J. Palmer proposed a 'duplex bridge' with different sliding sections that could open to allow tall-masted ships through. John Keith suggested a vast under-river arcade lined with shops, but this, along with a free ferry like the one later introduced by Sir Joseph Bazalgette at Woolwich, was vetoed immediately.

Sir Joseph Bazalgette was the Chief Engineer to the Metropolitan Board of Works. He rebuilt three of the older Thames bridges at Hammersmith, Putney and Battersea, and created London's sewage system. But his plan for a high-level bridge was deemed unsuitable by the parliamentary committee considering the proposals in 1878.

ABOVE: Design by Frederic Barnett for a 'duplex' low-level Tower Bridge, 1876.

A WATERMAN IN DOGGETTS COAT AND BADGE.

LEFT: A waterman in his splendid Doggett's coat and badge, *c.*1860.

'All I can say is bosh!' Objections to the bridge

The London watermen, who earned their crust carrying people across the Thames, were naturally not keen on a new bridge. Neither were some of the owners of the wharves, who feared a bridge would damage their trade.

Queen Victoria herself was among those most vexed by the prospect of a bridge near the Tower of London. Since Norman times the Tower had been an armoury and a famous gaol. Until 1810 it had also housed the Royal Mint, where the coins of the realm were struck.

It was an ancient totem of regal power and the Queen rejected any suggestion that a new river crossing would improve its standing, writing in a letter: 'to those who say the bridge will increase the defensive strength of the Tower and improve the beauty and historical associations of the place. All I can say is bosh!'

But a new bridge east of the Tower was just too important for London and even the Queen's objections were overruled.

BELOW: Panoramic view by Robert Havell of the River Thames with the Tower of London on the right, 1823.

The winning design

Sir Horace Jones was the Architect and Surveyor to the Corporation of London. He gave the city such architectural gems as Leadenhall, Smithfield and Billingsgate markets. Though a judge for the competition, he also submitted a plan of his own. Based on examples of lifting spans he had seen on the Continent, he put forward a scheme for a 'bascule' (French for 'seesaw') bridge.

His design featured a roadway in two parts that opened in the middle, each 'wing' pulled up like a drawbridge to allow ships to sail on into the Pool of London. The initial plan, which featured a great arch, was rejected by the committee, but Jones sought out John Wolfe Barry to improve his design.

The son of the famous architect Sir Charles Barry, who had designed the House of Parliament with Augustus Pugin, John Wolfe Barry was a pre-eminent builder of railway bridges, including those at Cannon Street and Blackfriars. Wolfe Barry's greater technical expertise proved just the ticket.

Jones and Wolfe Barry's new design, a bascule bridge with two Gothic towers and upper pedestrian walkways, was formally adopted with the passing of the Tower Bridge Act of 1885.

RIGHT: Sir Horace Jones's design for a bascule bridge, 1878.

Extension of Parapet N approach

Inch Scale

THE TOWER BRIDGE

THE MEMORIAL STONE WAS LAID BY H.R.H THE PRINCE OF WALES ON BEHALF OF H.M THE QUEEN ON 21ST JUNE 1886
THE RIGHT HONOURABLE JOHN STAPLES LORD MAYOR
EDWARD ATKINSON ESQ. CHAIRMAN OF THE BRIDGE HOUSE ESTATES COMMITTEE

JOHN WOLFE BARRY,
Engineer

SIR HORACE JONES,
Architect

ABOVE: Commemorative print of Tower Bridge, 1886, showing Horace Jones's projected design of the masonry in red brick.

Modern Gothic Something new and old

Tower Bridge was to use the most up-to-date technology of its age. Its bascules were to be powered by steam hydraulics and it was to be built in steel. And yet it could easily be mistaken for an ancient stone castle. Why make something new look so old?

The answer is that the bridge had to blend in with its surroundings. Conservationists were worried that the nearby Tower of London would be overshadowed by a shiny modern bridge. The design of Tower Bridge had to complement the older building. At this time it was fashionable to clad ordinary public buildings made of iron and steel in stone to give them the grandeur of cathedrals. The modern materials that made the structure of the bridge so cutting edge were therefore concealed from sight by Portland stone. One of the most striking new buildings in London in recent years, The Shard, offers a contemporary twist on much the same idea. Its architect Renzo Piano sought to emulate the spires of Sir Christopher Wren's City churches with his epic glass skyscraper.

ABOVE: Tower Bridge and The Shard.

LEFT: Plan by George Stevenson showing the east and west fronts of the towers, 1889.

Team Tower Bridge

271.00

248.00

241.25

223.00

214.25

Horace Jones

John Wolfe Barry

Horace Jones never lived to see Tower Bridge: he died quite suddenly on 21 May 1887, a day after his 68th birthday. But Jones and Wolfe Barry had assembled some of the finest engineers of their age to help design and build their bridge:

George Stevenson

Henry Marc Brunel

The second son of Isambard Kingdom Brunel, Henry Marc was the third and last of the Brunel engineering dynasty. The business partner of John Wolfe Barry, he was a gifted civil engineer with a love of the theatre, who provided many of the vital calculations and details for the construction of the bridge.

Formerly Horace Jones's assistant, George Stevenson worked tirelessly to refine the design of Tower Bridge. He was the man who gave the bridge its unique appearance, replacing the red-brick facing in the original plan with granite and Portland stonework, and adding cast-iron parapets and decorative panelling to the walkways.

Sir William Arrol

The son of a spinner, Arrol rose from working in a cotton mill to become one of the leading civil engineers and steel manufacturers in Britain. He made his name building the monumental Forth Bridge in Scotland, and supplied the steel for Tower Bridge.

Sir William Armstrong

Armstrong was an industrialist based in the northeast of England. His firm pioneered the development and manufacture of hydraulic cranes; it manufactured and installed all of the hydraulic equipment on Tower Bridge.

George Edward Wilson Cruttwell

Cruttwell held the position of Resident Engineer throughout the construction of the bridge and was a stalwart member of the team working on the ground.

Building the bridge

This mammoth operation started on 22 April 1886. John Wolfe Barry estimated it would take four years to complete at a cost of £585,000. But it actually took eight years to build and ended up costing a whopping £1,184,000.

In the course of the works, there were 29 serious injuries and 10 men died. Since the men worked with no safety netting and at heights of over 30 metres (100 feet), it is surprising there were not more deaths or injuries. Visibility was often poor. Before the passing of the Clean Air Act of 1956 London was regularly plagued by 'peasoupers': thick, yellow fogs produced by coal smoke and factory fumes. In January 1888 the *Daily News* reported that all work on the bridge had been suspended due to the arrival of a particularly bad fog.

The most arduous part of Tower Bridge's construction involved building the two foundation piers. At their completion in 1894 they were the largest bridge piers ever built. Each was sunk nearly 8 metres deep into the clay riverbed. The piers were constructed inside steel caissons – watertight enclosures that were pumped dry with the help of divers to allow work to continue far beneath the Thames.

To ensure Tower Bridge was sturdy enough, each of the two columns for the towers erected above the piers was built from four wrought-iron octagonal pillars of riveted plates and cross-braced together for extra strength. Ultimately 5 miles of steel plates were used for the columns on the piers and abutments. The abutments were also anchored to each shore with concrete. In the end, the bridge was strong enough to resist four times as much wind as legally required.

The steelwork was built in Glasgow and shipped to London on steamers. Portland stone from quarries in Dorset, granite from Cornwall and bricks made from Gault clay from Cambridgeshire and Bedfordshire were used to create the fabulous castle-type towers that perch at either end of the bridge. Welsh slate covered the roofs of the ornate turrets. Tower Bridge was a truly national landmark.

ABOVE: Progress on the bridge captured in old sepia photographs discovered decades later.

Up and down

The bridge moves on a pivot with a massive weight of 422 tons of lead and iron, counterbalancing the longer length of the other end. The counterweights permit the bascules to be raised to a nearly vertical 86-degree opening angle. The whole mechanism is operated by hydraulics; until 1976 steam was used to power two massive pumping engines and a further eight large hydraulic engines and six accumulators.

A signalling system was adopted so that anyone approaching the bridge would know the position of the bascules. Red indicated when the bridge was down or moving down; green when it was up or moving upwards. During daylight hours this was conveyed by semaphore flags and after dark by coloured lamps. In foggy conditions a gong was sounded instead.

The bridge is rarely ever opened to the full 86 degrees. Typical openings average from 20 to 50 degrees. The bridge has occasionally raised its bascules to mark great state events. These have included the return of Queen Elizabeth II from her first world tour on the Royal Yacht *Britannia* in 1954 and the funeral of the former war-time Prime Minister Sir Winston Churchill in 1965.

LEFT: A Port of London official striking a gong on Tower Bridge during foggy weather, 1955.

OPPOSITE: Coloured lantern slide capturing the image of Tower Bridge raised to let a steamer pass through, *c.*1895.

8.99 **metres**

Height of clearance when bridge closed

43.5 **metres**

Height of clearance when bridge open

The grand opening

Tower Bridge was opened with a lavish ceremony on Saturday 30 June 1894 by the Prince and Princess of Wales. At noon the royal couple led a grand procession of dignitaries in horse-drawn carriages across the bridge from the north to the south bank.

It was a bright summer day; spectators lined both banks of the river and stood on London Bridge. The Thames, glistening under the midday sun, thronged with gaily decorated boats and barges, and watermen ferried people eager for a closer view out on to the river in their wherries.

The royal cavalcade returned to the north bank, where the Prince of Wales was escorted to a pedestal, crowned with a silver cup that was rigged to trigger the hydraulics of the bridge. Declaring the bridge open for river traffic, the Prince

ABOVE: *The Opening Ceremony of the Tower Bridge* by William Wyllie, 1894–5.

For a moment, the great crowd hushed in silence. Then in a deafening shout of applause, which soared, as only a British cheer can soar, above the thunder of the Tower guns, above the ringing notes of the trumpets, and above the wild din from the sirens and the whistles of the steamers, they gave vent to their admiration and delight at the marvel they had been privileged to see. They had indeed witnessed a spectacle not easily forgotten. THE TIMES, 30 JUNE 1894

turned the cup and right on cue the mechanism sprang to life and started raising the bascules.

For the workers who had toiled so hard to create this mechanical marvel, there was a special celebratory meal in a marquee set up near the southern approach. Afterwards pipes, tobacco and boxes of sweets were distributed as mementoes of the occasion.

BELOW: Official souvenir programmes from the opening of Tower Bridge, 1894.

A colourful crossing

Tower Bridge was originally painted 'bright chocolate brown' – with three coats of paint in total required to cover the metalwork entirely. It was repainted roughly every seven years.

During the Second World War brown made way for battleship grey, this new livery intended to help camouflage the bridge from enemy planes making bombing raids on the capital.

In 1977, in honour of Queen Elizabeth II's silver jubilee, the patriotic colour scheme of red, white and blue it retains to this day was chosen.

Between 2009 and 2011, the whole bridge received a complete makeover, involving 22,000 litres of a specially developed type of hard-wearing paint to save on future maintenance costs.

22,000 litres
Amount of paint required to cover the bridge

LEFT: The patriotic red, white and blue colour scheme of the bridge.

RIGHT: Two men carrying out maintenance work on the bridge, 1957.

An engineering icon

Tower Bridge was built at the point when both photography and picture postcards were coming into their own. Daringly engineered bridges in iron and steel made ideal subjects for pioneering Victorian photographers armed with primitive plate-glass cameras.

Tower Bridge immediately became – and remains – one of London's most-photographed landmarks. The photo-sharing site Flickr currently lists over 600,000 images of Tower Bridge on its database.

But back when only the wealthy could afford cameras, picture postcards offered a cheap and easy alternative to a photograph of your own. These cards carried the bridge's likeness around the globe and helped turn it into an iconic symbol of London.

As the largest and most sophisticated bascule bridge ever built, Tower Bridge was soon nicknamed 'The Wonder Bridge'. It quickly won the hearts of Londoners, especially many of the poorer residents of southeast London who had previously been forced to trek to London Bridge to make their journeys north for factory or warehouse work on the other bank.

ABOVE: Postcards of Tower Bridge, c.1920s (left) and 1910 (right).

LEFT: Poster advertising
Air France flights to London.

RIGHT: Advertisement for
KLM flights with iconic
symbols of European cities.

A working landmark

During its first year the bridge opened 6,160 times. Some 8,000 horse-drawn vehicles and 60,000 pedestrians crossed it daily. It opened up 20 or 30 times a day and the boilers for the steam to operate the mechanism ate up 20 tons of coal every week. Nowadays it is raised electronically and only three to four times a day.

Originally some 80 staff were employed to operate and maintain the bridge under the command of the bridge master and a resident engineer, both of whom lived on site. The very first Tower Bridge Master was Lieutenant Bertie Angelo Cator. Cator was appointed on an annual salary of £300 and took up his duties on 14 May 1894, just six weeks before the opening. He, or his deputy, had to be on duty day and night, together with engine drivers, watchmen, signalmen, firemen and eight constables from the City of London police. A tugboat was also permanently on hand to direct river traffic and to help any ships in difficulty.

The resident engineer was responsible for ensuring that the bridge's hydraulic machinery was in good order and that any damage caused through wear and tear and accidents was repaired, such as when ships collided with its piers.

Under him was a dedicated maintenance crew of 30 men from the bridge's staff. There were blacksmiths, carpenters and plumbers, each with their own workshops. The blacksmiths also looked after the bridge's horses.

To prevent any delays to traffic, a team of reserve horses was kept constantly on standby in a stable under the southern approach in case a dray collapsed while leading a cart across the bridge.

BELOW: Lighters approaching the General Steam Navigation Company's Wharf next to Tower Bridge, c.1905.

18 tons

Weight limit for vehicles using the bridge

SPEED LIMIT
VEHICLES WITH OR
WITHOUT TRAILERS
OVER 12 TONS
10 M.P.H.
OTHER VEHICLES
20 M.P.H.

20 mph

Speed limit for vehicles using the bridge

ABOVE: Photo of a mist-covered Tower Bridge by Edward Charles Norrington, c.1959.

LEFT: Postcard showing traffic on Tower Bridge, c.1910.

56 LONDON. — The Tower Bridge. — Looking North. — LL.

BELOW: The upper walkways between the north and south towers.

OPPOSITE: The bridge brilliantly illuminated at night.

90 seconds

How long it takes to raise the bascules

Walking on water

When the bridge was open for ships, pedestrians could either wait for the bascules to rise and fall, or go to the top of the towers and use the two high walkways. To reach them involved climbing up 206 steps. Fortunately there were also lifts powered by the same hydraulics that operated the bascules. These could carry 25 people and made 25 trips up and down every hour.

The walkways opened at sunrise and, if gas-lit, closed each night at sunset, offering magnificent panoramic views of London. But they became notorious as a haunt of pickpockets and prostitutes, and were closed to the public in 1910. They would not be opened again until 1982.

One of the less pleasant duties the bridge staff had to perform was fishing dead bodies out of the Thames. There were victims of murder and those who had drowned or committed suicide, but many were poor Londoners whose families could not afford the cost of a burial, so their bodies were never reclaimed.

Once retrieved from the water, the corpses were placed in a tiled vault in the north pier until they could be removed to a mortuary. This area of the bridge soon acquired the rather grisly nickname 'Dead Man's Hole'.

The Tower Bridge pleasure beach

In 1934 an artificial beach was opened on the foreshore next to Tower Bridge; it was finally closed in 1971 due to fears about pollution.

In the late Victorian era diving became an exciting new gymnastic sport. 'Aquatic performers' known as 'Professors' entertained holidaymakers at coastal resorts by diving off seaside piers. In London, jumping off the bridges over the Thames was against the law, but this didn't stop daredevil divers staging illicit stunts.

In 1899 the 17-year-old 'Lady Swimmer' Marie Finney made the news by leaping off London Bridge. But her brother, William Finney, would top that on 12 April 1901 by becoming the first person ever to perform a 'sensational dive' from Tower Bridge.

The Thames was declared biologically dead in 1957. But a drastic reduction in pollution in recent years means that today the river boasts over 125 species of fish. Its foreshore is again in use as a spot for picnics, sunbathing and mudlarking.

LEFT: A beefeater helps to build sandcastles on Tower Beach, 1950s.

OPPOSITE: Londoners relax on Tower Beach, 1952.

Lighting up London

Tower Bridge has always been one of the brightest features of the capital's skyline. When it first opened it was lit by 200 of the latest gas lamps supplied by the London firm William Sugg and Co., built at their Vincent Works factory in Westminster. It remained gas-lit until 1966 when electric lamps were installed.

Since 2012 Tower Bridge has been lit by more environmentally friendly LED ('light-emitting diode') lighting. This uses 40% less power than its previous lighting system and can emit a range of different colours. The lights were first switched on for the Queen's diamond jubilee and used to great effect during the 2012 Olympic Games, when the crossing was also fitted with giant Olympic rings and Paralympic agitos for the duration of the sporting contest. Every evening that a member of Team GB won a gold medal, the bridge also glowed gold, cheering each home win with its dazzling new lights.

ABOVE: Tower Bridge shows off its new multicoloured lights.

OPPOSITE: Fireworks for the Olympics opening ceremony, 2012.

40,000

Average number of people
to cross the bridge each day

ABOVE: Irish aviator Frank
McClean flies his Short
seaplane through Tower
Bridge, 10 August 1912.

LEFT: Frank McClean at
Eastchurch, 1910.

The bridge takes flight

ABOVE: Major Thomas Hans Orde-Lees captured making the first parachute jump from the bridge, 1917.

RIGHT: An artist's impression of Alan Pollock's RAF Hunter jet flying under the walkway, 1968.

Born in England to a wealthy Irish family, Frank McClean was a pioneering aviator. He flew some of the earliest planes ever built by the Wright brothers in France in 1908. He was also a bit of a show-off: on 10 August 1912 he stunned spectators and newspaper reporters by flying his Short pusher seaplane between the bascules and the high-level walkways of Tower Bridge.

In 1917 Thomas Hans Orde-Lees, a member of the Royal Flying Corps, the forerunner of the RAF, conducted a parachute jump off Tower Bridge. Parachutes were then in their infancy, and Lees was attempting to convince the British military of their effectiveness. Known as the 'Mad Major' and a member of Ernest Shackleton's Antarctic expedition, Orde-Lees's Tower Bridge jump is thought to be the lowest voluntary parachute descent ever made. But it seemed to do the trick. The Royal Flying Corps promptly formed a parachute division and appointed him its commander-in-chief.

The Royal Air Force celebrated its 50th anniversary in 1968 and held a golden jubilee dinner attended by the Queen. But many were disappointed that officials had blocked a previously promised ceremonial fly-past over London. Alan Richard Pollock, a 32-year-old senior flight commander at West Raynham, chose to protest against this decision by putting on an air display of his own. On 5 April he flew an RAF Hunter jet past the Houses of Parliament and then on under Tower Bridge's walkway at a speed of 400 miles per hour. He was placed under arrest upon landing, but later discharged from the RAF on medical grounds.

The bridge at war

With the outbreak of the First World War in 1914 London soon found itself under attack from Zeppelin raids and the whole country was threatened by a German naval invasion from the North Sea. To protect the city, Royal Marines were deployed on Tower Bridge. Later a gunnery battalion, manned mostly by veterans who had served in France and Belgium, was also stationed there.

As such a famous London landmark, Tower Bridge became an obvious target for enemy fighter planes in World War Two. Virtually a sitting duck, it came close to being burnt to a crisp on the opening night of the Blitz on 7 September 1940.

By 6.30 that evening, nine fires were reported as 'out of control in the docks' and a dockside rum warehouse had been bombed, spilling its contents into the water and setting the Thames alight 'like a Christmas pudding'.

Despite this, Tower Bridge survived the war relatively unscathed. The neighbouring St Katherine's Docks and the area to the northeast of St Paul's Cathedral were obliterated in the subsequent raids.

There were 17 attacks on the Houses of Parliament and the Tower of London suffered 15 direct hits. Tower Bridge, by contrast, suffered only minor damage to its windows, roof slates and stone cladding.

A bomb did glance the south shore span in another raid on 16 April 1941, when five people where injured by flying glass and debris. One of the tugs used to direct river traffic was, however, destroyed by a V-1 flying bomb or doodlebug in the closing year of the war.

Much like St Paul's Cathedral, Tower Bridge's survival came to represent the unwavering determination of London through the bitterest days of the war.

ABOVE: Enemy aircraft over London as depicted by a German war artist, 1940.

OPPOSITE: *Blitz. Our London Docks* by Charles Pears, 1940.

The bus that didn't stop

On 30 December 1952 a number 78 double-decker bus had to make a daring leap over a 3-foot gap, after a watchman failed to ring the bell to announce that the bridge was opening.

The driver Albert Gunter told *The Times* newspaper: 'I had to keep going, otherwise we should have been in the water. I suddenly saw the road in front of me appeared to be sinking.'

You can still catch a number 78 bus over Tower Bridge, but automated traffic lights and gates make it impossible for accidents like that to happen today.

LEFT: A modern-day double-decker bus travelling over Tower Bridge.

BELOW: Albert Gunter, the driver behind the wheel at the time of the daring bus leap.

OPPOSITE: *Now or Never*, illustration of the bus jump by Rachel Hunt, 2014.

The changing Thames

By the start of the 1960s traffic on the Thames had gone into sharp decline, with freight increasingly now transported by road as new motorways started to be built across Britain. From April 1962, and in response to the falling number of boats, Tower Bridge began to require all ships' captains to give 24 hours' notice if they wanted the bascules raised.

Worse was to come for the old working river with the advent of containerised shipping. With the exception of Tilbury, some 24 miles downstream of Tower Bridge, the capital's docks were simply too small to accommodate the new larger container ships whose cargo could be loaded and unloaded with a much smaller workforce. Between 1967 and 1981, and beginning with the East India Docks and ending with the Royals, all of London's docks were to close with the loss of tens of thousands of jobs. In the aftermath 8 square miles of dockland warehouses and wharves were left deserted.

OPPOSITE: Boats and barges line the foreshore by Tower Bridge, 1959.

BELOW: Tower Bridge at sunrise, 2010.

Running out of steam

New technology would eventually reduce the number of people needed to operate the bridge. In 1972 the steam-driven power engines which had served the bridge without fail for 80 years were replaced by a new electric oil-hydraulic system. Five years later new electric controls, which raised the bascules at the push of button, were installed and the bridge could be run with a staff of a dozen. Gone, too, were the days of coal, smoke and heat. The bridge was now opening on average only seven or eight times a week.

If river traffic was falling, road use continued to soar with some 10,000 commercial vehicles thundering across Tower Bridge every single day by the late 1970s. Huge juggernauts and container lorries, often weighing in excess of 30 tons each, were a far cry from the ambling horse-drawn carts of Victorian days. Amid concerns about the damage they might be doing to the bridge, the Greater London Council imposed a 5-ton limit on unladen vehicles from the end of January 1979.

RIGHT AND FAR RIGHT: The Engine Room, once the beating heart of Tower Bridge.

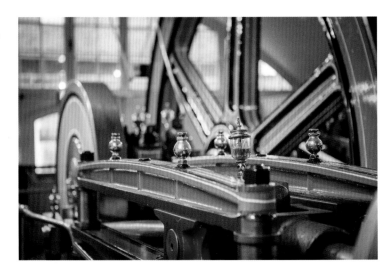

1,000

Approx. number of times the bascules are raised each year

TOP: Stoker Fred Woodward in the boiler room in 1967, when the bridge was still operated by steam power.

ABOVE: Tom Mascall operating the levers of machinery in the control room in 1967. Although built in 1894, they were still functioning faultlessly.

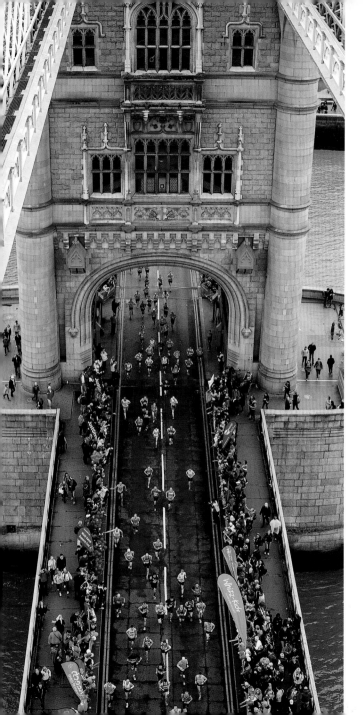

New life for an old bridge

On 30 June 1982 a new chapter in the life of Tower Bridge began when the upper walkways, freshly encased in glass-reinforced plastic, were reopened to the public for the first time since 1910.

This was part of an extensive renovation programme that also saw a brand new underground exhibition centre created on the south bank of the bridge. Here visitors could finally see the Victorian machinery that had run the bridge so reliably over the years. If turning these once vital engines into museum pieces seems a bit backward-looking, Tower Bridge was actually being rather more prescient than nostalgic. For in the coming decade Britain would move from being an industrial economy to one based around services and finance.

On Tower Bridge's own doorstep all those decayed wharves and warehouses would gradually be reinvented as shopping centres, restaurants, bars, art galleries and theatres, while a whole new financial district, Canary Wharf, arose on the site of the old West India Docks on the Isle of Dogs.

43.5 metres
Height of walkways over water at High Tide

LEFT: Visitors enjoy the walkways that were closed to the public for 72 years.

OPPOSITE: A view of the bridge roadway from above during the London Marathon, 2015.

530 kilograms

Weight of each glass floor panel

A bridge for London, yesterday, today and tomorrow

London is still a trading city. The commodities bought and sold may no longer enter London's docks physically, but the money is handled by City institutions virtually, and trade of all stripes continues to make the capital such a vibrant place. With the redevelopment of the old docklands, Tower Bridge has become once again a gateway to the financial heart of London.

Equally, the Thames has been restored to the centre of city life, and river traffic is up, if mostly for pleasure rather than business, with the bridge now raising its bascules about a thousand times a year.

No longer merely a river crossing, Tower Bridge in the 21st century is an events space, a concert hall, an education centre and, as one of the most photographed landmarks in the world, a hugely popular location for film and television crews making blockbuster movies and hit shows.

In so many respects, Tower Bridge is the perfect emblem for London, an ever-changing city that has ceaselessly adapted to altered circumstances and continues to thrive today.

BELOW: Tower Bridge and the southern river front as it looks today.